WE ARE PLEIADIAN

OUR STORY AS TOLD BY PLEIADIAN MASTERS

PLEIADIAN SAYING: "HE WHO GOES WITH THE FLOW IS ALWAYS THE LUCKIEST MAN STANDING." - JAMES CARWIN

GOLDEN SPIRAL PRESS

JANIE JURKOVICH

GOLDEN SPIRAL PRESS

We Are Pleiadian

Our Story as told by Pleiadian Masters

and channeled by Janie Jurkovich

Golden Spiral Press ™ is a trademark of Janie Jurkovich.

Published by Golden Spiral Press, Inc.

Clovis CA 93611 www.goldenspiralpress.net

Author Photos © Blanca Madrigal

Cover Image by Ellie Dote

1st Edition

ISBN: ISBN: 978-1-7369476-6-1

Table Of Contents

Introduction

As a spirit medium I can connect to beings in other dimensions almost effortlessly. My last book, The New I AM Document, Volume I, was entirely downloaded from Archangels. It was like I was dictating their words, but the words were actually felt versus being heard.

When I was approached in 2020 by the Pleiadians to transcribe their book, I found it interesting because I had never thought about the process of extraterrestrial beings coming to Earth and serendipitously joining our society. I transcribed a couple chapters and then lamented to them something along the lines of, "I'm busy doing other work. Do I have to do this?" They replied that I didn't, so I was happy to have one less task on my to do list.

One day I mentioned this seemingly strange occurrence to a friend. She replied that I really did need to transcribe their book because humanity is changing, and the world needs to hear that message. Ok; I told her I'd think about it. But she kept prodding, so I transcribed a bit more and sent her the pages. Eventually I embraced the process even though it was a bit outside my comfort zone.

It's one thing to admit you talk to deceased family members or even famous people who have passed. But telling the world that you also connect with extraterrestrials can be a bit hard to fathom.

I suppose I was wondering how I was going to explain talking to ET's to family, friends, and my marketing folks. Nonetheless, I plugged along, and it turned out to be a rather intriguing story.

I felt a sense of accomplishment because I had been told that I am part Pleiadian. Now that was quite something to embrace. It's one thing to know you are part Scottish, Irish, English, or from the former Yugoslavia, but telling others that your ancestors come from another planet, is way out of my comfort zone!

Recently I attended a GAIA channeling event in beautiful Louisville, Colorado. The last channeler, Darryl Anka who channels Bashar, was adamant that society will become fully aware of ET's within the next five years! That was a shocker to me; not that they exist, but **the fact that they exist will become undeniable** even to mainstream folks.

So I surmise that the timing is right and like my friend Michelle said, I need to get the story of the Pleiadians out. Enjoy this fascinating journey!

The Beginning

A long time ago in a faraway galaxy existed a people known as Pleiadians. Their numbers were few, but their vision was mighty. They were space travelers or extraterrestrials as we now know them.

Their small planet was riddled with wars, famine, and darkness, much like we've experienced here on Earth.

However, the people didn't want their planet to die. They wanted to keep living and provide a safe place for their offspring. They traveled from place to place till they finally settled on the great planet Earth.

They were happy and all was well.

The Escape

They gathered in the starship and took one last look at their home before they bid it goodbye. Pleiades, their home, was ravaged by the effects of war, fire, and disease.

They were off to find new lands where they could procreate with others and live a happy life. They were sad to go but knew the future of their species depended on it. They did not want to succumb to extinction like so many of their neighbors had, namely the Lyrans. Luckily, a few had fled in time and repopulated with the Pleiades. These were the ones on the starship now.

A great new race who had risen from the ashes as it were and they were willing to do what it took to maintain their race, despite the hardships, despite the odds. Their homeland was no longer an option with all the death and destruction. They were willing to do the work, take the chances and answer the call of the wild.

Little did they know where it would lead them. Little did they know that yes, at the time, their very existence as a species would depend on their brave actions...the actions that led to this escape from a fiery planet in the northern hemisphere called Seven Sisters. A life-changing moment for all the Universe.

For you see, this was the first step in obtaining their freedom, taking control of their lives, something we all must learn how to do at some point.

Take charge. Take action. Make good on those plans. Make our marks and that of society's. It was a start and later they would realize just what their bravery had cost them and where it had led them…on an adventure of a lifetime. And it all started eons ago.

The story unravels.

The Mission

The reason for our mission was to be able to sustain our culture and our very lives.

Sure, we could have remained and succumbed to famine and wars, but we were doubtful about a sustained society. It was as if we had taken all we could from our small planet. Surely a lesson to be learned here on Earth.

When we settled on Earth, it was a long time ago before the records of man. We helped build a civilized society. Our looks adapted to your situation here. We didn't originally look like people do now – even as varied as they are.

We were blue and taller, much like those in Avatar. It's the reason the movie was made; it was a replica of us. You see, there are many of us and some "remember." Movies and such are their ways to start to teach others about our heritage.

So, when we came, ours was a great civilization (at least it had been at one time) much more advanced than human civilizations are on Earth now.

We had great telepathic abilities, good nurturing of children, and a better connectedness to each other. It took us eons to get where we are now. Our progress may seem slow, but you people (meaning other humans here at this time) are very slow learners, at least to our way of thinking.

It was a great lesson to us that humankind on Earth would take so long in our eyes to evolve, although humanity is making good progress now.

Let us continue. We had to start from scratch and infiltrate your society one by one. We went from observing to "becoming" one of you. (We are using the term 'you,' but we do understand that we are all One and that you are Pleiadian by birth.) Anyway, it took a long time for us to assimilate into human culture on Earth and become one with you because it occurred in the birth process.

One of "ours" (the Pleiadians) would join the soul of the unborn and replace it as it were with our soul. You see, we are really spirit—an unseen vibration of light. We were merely blue and tall as a way to live or exist on our previous planet, just like you are on Earth.

Because of our advanced nature, we were able to do this.

You keep wondering about Elon Musk and Nicola Tesla. Yes, they were or are Pleiadians. This is the reason they could move the planet with their forward thinking. Their spirits brought those lessons with them. They knew or know certain things are possible, so they strove to create them on Earth. They "know" it can be done. They might not remember or know "why" but they believe it to be so.

This is the reason we should all pay attention to our intuition (our guidance system) for it is our way to know to learn from the past and sometimes the future.

You are part of a great race, a race that will someday take over Earth with its loving ways. Do not squander your heritage for it is a true blessing.

We love you. Thank you for the mission.

Your Pleiadian Masters

Mastery of the Process

Mastery of the process of Pleiadians infiltrating humankind has been slow and fruitful.

We are happy that so many of you present on the planet Earth right now are willing to accept the challenge.

The goal is to teach other beings how to live a happy and blissful life without so much pain and sorrow. You see, although pain and sorrow are indeed part of the human experience, it doesn't have to be so all-inclusive as some beings seem to think.

Our creator, the Great One, wants us to be happy, to exist with a spring in our steps and a big smile on our faces. There is no need for worry, shame, and hatred. These are some of the traits and experiences we Pleiadians wish to extoll to the masses.

Although the lessons may seem simple, we understand that adopting them to one's everyday life will take time. Humans sometimes need to experience their lessons repeatedly to really know them or "absorb them into their souls," as you say.

We wish for all humans to experience and follow these lessons. Our work here on Earth will continue until all beings, currently in human form, learn and exemplify these teachings. Then our work will be done.

However, as you might imagine, this is a huge undertaking, one that will take lifetimes to achieve. Each person's part, however small, contributes to the attainment of the joint mission. Your part of transcribing this document (for which we thank you greatly) is but a small part of the mission.

Others are working diligently on their parts as you have seen and will continue to see in future weeks and years.

All beings doing their part will contribute to a significant shift in the way life is experienced and lived on Earth.

One way, in fact the first way, is for humans to be more open. They must be more open to change, possibilities, and a way to look at and experience human life.

Many are stuck on the drudgeries of human existence...work, family, food, and shelter. There is so much more to life than this! Learning to appreciate life and show gratitude for what you've been given or attained, is essential for moving forward.

For you see, once our basic human needs are cared for (which shouldn't be nearly as difficult in current times to accomplish as many seem to believe) then humankind can work on the other levels – the unseen ones.

Tactics like dealing with negative people, learning to turn challenges into opportunities, helping raise others up to a higher level of consciousness, and actively pursuing one's mission can be accomplished.

Life is merely a mirage of lessons to be completed prior to reaching these new levels of understanding.

If each being does his or her part, we will reach this new level of understanding for all humanity, which is our ultimate goal.

Words of Wisdom

We, the Pleiadian masters, offer these words of wisdom:

-Do not covet **money** for yourself. Instead, use it to bring happiness and goodwill to others. In turn, you will find yourself happier, for the energy of money when properly used can heal many a heartache or loss. (Not all heartaches are lovers; many are things you feel lacking in your life!) Money is nothing more than energy and when shared with others, brings about a comradery or shared happiness. Something to think about when deciding to spend money on something that benefits another person and not so much yourself.

-Another teaching revolves around **time**, for it is endless, has no beginning and has no end, but fail to use it when the time comes, and it will be lost forever. Some may refer to this as "lost opportunities." You see, sometimes the Universe (those of us working behind the scenes conspiring to bring your manifestations into being) will set up a complicated set of circumstances to bring about your wishes. However, many of you squander these actions (complicated as they are!) by refusing to take the next step that will in fact, bring your manifestation into action. We (those beings in another realm) can only do so much. The rest is up to you.

Please, we encourage you to take advantage of these synchronicities before they become "lost opportunities." You'll never know where these roads will take you if you don't take advantage of them.

-Another tactic or area for improvement we often see relates to **love**. Why are you humans so stingy (for lack of a better word) with your love? Love is merely a feeling, a burst of energy that keeps the life force of another going. It's merely a light that can keep another shining for if a person doesn't feel love of some sort while on their journey (human or animal love, such as from a pet) they will surely die.

Lack of love or being loved, causes a soul to wither and die! It's like having a plant and refusing to water it or offer sunshine. Love is a requirement for human existence. You don't need to love everyone in a sexual manner. This is not what we mean.

You don't need to include hateful or mean people in your life either. Just love them for who they are and where they are on their Earthly journey. "Love them from a distance," might be the way to consider such situations. For once they feel the love from you and others, they will shift. They will change and become happier and more loving people in the process. This method can be equaled to God's commandment to "Love one another." This is the reason he wrote it.

The love in each of us will heal humanity. In fact, it is the only thing that will. Other methods are merely pathways to teaching humans how to properly love one and another. The greatest blessing of all is to learn to love everyone and everything despite the many differences. The differences are what make each of us unique. Learn to love the differences or at least accept the differences and send loving thoughts to one another.

Such actions, when multiplied tenfold, hundred-fold, and so forth, will make all the difference. It will bring humanity closer. It will raise the joint consciousness of humanity. A truly enormous feat and one that is required for advancement.

Family Matters

Once again we discuss the many challenges facing humankind and how to deal with them.

Today our lesson will be regarding family members. There are two types of families–those you're born with and those you create.

You see, some of you beings on Earth fall into a happy, joyous, and complete family with few if any challenges to conquer on your life's journey. Most beings, however, do not. Their journey is often fraught with complicated relationships bringing a multitude of issues to deal with. Learning to deal with these relational issues may be the crux of one's life mission for it is that complicated.

For those willing souls, there are more lessons to learn at the next level–with other people, things, and circumstances outside your personal family. These folks often learn to adopt their own "new" family that will in turn help them to advance sooner. Perhaps their "born" family is not capable of advancing this go-round in life. Perhaps their "born" or original family has too many self-inflicted issues to manage so that they cannot support anyone else emotionally, financially, or spiritually. It could be their mission this time was much more limited than the path you or other family member need to follow.

Nonetheless, the necessity often arises to reach out and find or collect your own proper family–a group of beings who can offer mutual support and well-being so all members can grow and learn, hence advancing as it is required for those seeking enlightenment.

Sometimes finding the right mix of souls to join you on your path can be a conundrum, but do not despair, for if it were easy, what lessons would you learn? Few, if any.

So, one's journey to find your new family, often called a soul family, will be fraught with many lessons, most of which are about interpersonal relationships. These include the value placed on money, one's sexual proclivities, trust, understanding, support, honesty, and a host of other more emotionally based actions or traits. The emphasis is on learning how to interact with others whilst learning these many lessons.

The key to navigating these scenarios already exists inside each and every one of you. The "test" as it were, is learning to access this internal mechanism for mutual good for self and others.

This internal guidance system is your heart (or soul) as some may refer to it. It is located in your upper chest. It is activated or working when one senses "feelings"– bad, good, or maybe ambivalent feelings. Learning to notice and then interpret these "feelings" from the heart chakra will enable one to advance more quicky on the time line of one's life journey.

Many beings unfortunately tend to ignore these inklings or small feelings, pushing ahead with life by ignoring them.

We implore you not to do this! This is a God-given system designed to help you. So please don't ignore it! Instead learn to be quiet and slow down. Learn to feel the differences. Memorize the different feelings and the messages associated with them. Learn to interpret their meanings. Lastly, learn to act according to this sensory system you've been granted with, for it will make your journey so much more pleasant. Additionally, you'll become quite adept at accessing your guidance system and therefore your advancement will tend to be more seamless and quicker. At any rate, learning to access your feeling state and following it will certainly feel better in your soul.

This system can be used to develop ways to distance you from fraught relationships that you will never be able to heal. It will help you develop new relationships with others, hopefully beings from which you will both learn lessons and grow for a time, even if it doesn't remain a long relationship.

At least you'll be learning. At least you'll be growing. These are the keys to progress on your Earthly adventure.

Later as you become more adept at accessing this system, you'll be able to recognize your soul family members. It may "feel" somewhat like a homecoming or recognition of an old friend. These are relationships to treasure, for finding a soul family member on an Earthly journey is a quite remarkable occurrence.

However, your guidance system can help you locate and recognize who you should allow to be in your new soul family. Remember, they should all help each other grow and learn, thus advancing. This means at times, lessons will be tough because you must hold each other accountable, which is sometimes not easy to do.

You will undoubtedly make some mistakes and so will they but try to remember that life is a journey more like a process that is honed by time and experience. Think of you and your soul family as a group being held in a testing lab of sorts, where your interactions are being observed by others so they can learn. Try to have some fun with this venture too, for life is meant to be enjoyed, not a worrisome event.

Learning to properly access and follow your internal guidance system will help you immensely on your journey. Use it to locate your soul family or tribe as some may call it, so that as a group you can make the most advancement.

Diplomatically Speaking

Now we will address the process or learned behavior of becoming a diplomat.

We use this term to describe having a good relationship with other beings in a compatible way. As you know, some beings can be most difficult to have a good relationship with, whether it's in your family or a business setting. How do we manage these relationships while continuing to grow on our own path?

First of all, if the circumstances are extremely tense or life-threatening, we must seek safety for our own well-being.

Most interactions aren't quite so severe (luckily), but we still must learn to distance ourselves from these obstacles. (We refer to these other beings as "obstacles" because that's what they are–obstacles to our growth and that of humanity.) By distancing ourselves from these sometimes nasty and outrageous individuals, we are protecting our own well-being much like one protects itself from an imminent danger like a train coming at you head on. Protecting our mind is just as important as protecting our body. (Think about that statement for a minute!)

So, the question becomes, how do we separate or remove ourselves from the presence of these less-than-stellar beings?

One way is to simply excuse yourself and move on. That method may not always work so more forceful methods may be necessary. In such situations, thank the person for whatever words of wisdom they have given you and then move on.

If they remain squarely in your face and you still feel uneasy, just be quiet and let your energy force (your life force) protect you. Silently call upon it to protect you from the negative energy being dispensed by the culprit. Then, once their tirade diminishes, you can walk away and return to your happy life without interruption or further annoyance.

The key is recognizing the severity of the situation early on so as to protect yourself with your own internal shield.

Your guidance system is how you do this. You will feel uneasy, and the feeling will grow! Perhaps your tummy will ache. Perhaps you may shake. Perhaps the color will drain from your face. Be aware of your symptoms and instill your own protective shield until you can diplomatically plan your escape!

You may wonder, how is this other being going to learn to correct their ways if you don't correct them?

Well, that is not your job. Your job is to learn to tolerate their behavior as best you can. They will continue to learn their lessons as they go through life with their terrible negative energy field and have interactions with others. Others will teach this being some lessons by the manner of their interactions. Each person's reaction will slowly affect the negative being. They will eventually change through various methods–whether it be by confrontation or the deafening quietness they experience.

Remember your job is you, first of all. Then your job is to do your part and your part alone of influencing others. You don't have to "fix" them or set them straight! This will come in time with the help of other beings.

This is how we all learn. In fact, this is how you learned too. Whether it was during this lifetime or the many previous ones. We all learn by practice, just as one learns to read a book or ride a bicycle.

Such is the cycle of life. We learn from each other, constantly changing and growing until we become diplomats as it were.

All for today. Tomorrow, more about diplomatic procedures.

Diplomatic Procedures

Whilst there are many types of personal interrelations and it is not possible to address every type, we will attempt to describe a few more situations along with our suggested methods for managing them.

Let's say you come across a rather gruff codgety older man. He seems set in his ways. He appears to be grumpy or a bit unhappy. He is not friendly at all. But you will find, if you address him politely and with kindness and patience, he may very well extend his personality outside the seemingly shell from which he exists.

The reason? He has self-manufactured his shell (physically and mentally) as a protective mechanism to keep the outside world out and keep himself safely locked in his comfort zone.

Your kindling of a small conversation can very well draw him out, much like a turtle who sticks his neck out of his shell to grab an insect or a piece of grass. Continued peaceful interactions may very well open him up to future conversations with others, thus changing or transforming his entire demeanor!

That is how it occurs. You play your part as a kind diplomat. Others continue the play by adding their parts and the man becomes a jovial older and wiser man, perhaps even well-loved by his social circle.

We all have some powers if we use them diplomatically.

Now, take a sullen and withdrawn young woman. She has led a tattered, unloved, and perhaps abused life so far. These circumstances have left her scared and terribly withdrawn. She is unable to deal with life on a day-to-day basis because she is fretful about all sorts of things. Her worry about being accosted. Her worry about a beating. Her worry of not being accepted. Each worry compounds on top of another and she is a bundle of nerves. She cannot, therefore, interact with others in even the most basic of ways. Her fears literally have her paralyzed.

But with just a few acts of kindness from another (perhaps you?), a little smile, a tender touch, or other helpful gesture, she can start the slow path of trust. She will learn to trust other beings, other circumstances, and trust that her future will be brighter. It's a process, but such transformation can take place if others do their parts to help heal her pain and gently show her other ways to live. It is possible if we all act as diplomats.

Now, take that boisterous child, filled with curiosity and energy. He wants to explore the highest part of the tree. He wants to build the largest, most elaborate building with his blocks. He seeks to design the most preposterous system or gadget with the toys and household items he has available to him.

Now if he is met with parents who shun his imagination and scold him for creating such a mess, how long will it be before his imagination and drive disappear and he learns to sit quietly in a corner? How long before he becomes stagnant–neither growing nor learning–let alone contributing to society?

You see, we can influence others in negative ways too. Consider this: his parents fuel (or at least play along with) his imagination. In future years perhaps he will create a great invention or solve some mathematical or structural quandary.

There is a difference between being diplomatic and merely so-so in a relationship. The diplomat helps guide others to becoming more of themselves, not more of what society desires. It's a way of spurring growth and self-actualization, helping others to become the best version of themselves. Help others to step into greatness, for we all have this greatness inside of us, waiting to be exposed so we can live our true lives–the one that feels right in our soul and complements the rest of humanity.

We hope you can see now why becoming a diplomat is helpful to others and to you in the long run, as we all travel the road to ascension (or enlightenment) together.

Hearsay

Today we'll discuss hearsay and the proper way to handle it. We (meaning all humans) do this. It can be a good thing or a not-so-good thing.

Hearsay is when you hear a story, event or situation and you repeat it to another, often embellishing it or exaggerating the event in an attempt to make the story-telling more interesting. The person may or may not realize just how dangerous this practice can be, so we want to shed some light upon this process.

Often the person who originates the statement doesn't have the full facts about the event. Remember they are only relating the event through their eyes, often without considering other related factors. You could say their view is somewhat tainted. They aren't doing this on purpose. It's just their point of view.

The person who hears their recantation most likely will put their own spin on the event, adding or emphasizing the portion of the story most meaningful to them. For instance, if the event includes a small part about bees and they are deathly afraid or allergic to bees, then that part of the story gains emphasis or importance. To the original storyteller, bees are merely a side fact of little importance.

Do you see how one's perception of an event just naturally shapes the story? Do you see how stories can become exaggerated, even to the point of being totally false? This is the quandary of hearsay.

That is not to say that hearsay is not important or has no value, for it certainly does. Think about the situation in early life on Earth. People would naturally want to warn others in their tribe about upcoming dangers like wild animals or treacherous terrain.

In times of cowboys and Indians one person known as the scout would ride ahead and return to the pack with news of any dangers. One would refer to this transfer of knowledge as hearsay.

Now let's say the scout returned to his group with news of an upcoming danger. Then the leader would decide perhaps a new course of action, depending on the knowledge presented. You can only imagine if the hearsay data were not accurate, how the actions of the group could be too severe or not severe enough.

Visions come to mind of the following two scenarios. When the scout returns to say, "Indians are among us!" Perhaps it's only a squaw woman with her children gathering berries (who can easily be avoided) in contrast to a full battalion of warrior-ready men. Quite a difference!

This is why it's important not to embellish hearsay, especially for one's own purpose. It might be an entertaining story with those embellishments, but exuding such a story as fact versus entertainment can lead to unwelcome results.

We want to encourage you to think about the circumstances when both hearing and relating hearsay and be cautious about the results your action may elicit.

Trails

Today's lesson will be quite a contrast. It's all about Trails–trails we leave behind and new trails we blaze.

You see when each of us comes into human existence, we leave behind a tiny part of ourselves to filter through the Universe as it were. This little part of us is known as a trail, a little piece of us left behind for all eternity.

As we go through our human incarnations, we develop new skills and abilities. We then use these skills to fabricate new trails that we once again leave behind when the time comes.

How does this occur, you may wonder? It might even seem beyond human capabilities, but it is not. The Great One bestowed this ability on each one of us, along with other abilities, most of which are still unknown to humanity.

As we learn and grow, hopefully we will come into contact with this ability and learn to access it for our own good and for the benefit of humanity.

Let us give an example. Let's say a being from Pleiadian heritage comes into life as a human being. He or she is lovingly nurtured by the parents without much unusual or extraordinary turmoil. As they gain maturity, they start to recognize their psychic abilities since initiation practices were common in their family. It's not such a stretch or shock that they possess this ability. They go forth in their younger adult formative years, interacting with other young folks, explaining, and sharing their ability. Their social circle then grows up thinking about intuition and psychic ability as a normal part of life, much like others might be indoctrinated into certain political parties or religions, beliefs held by their friends or family members.

This is how one leaves a trail...a little bit of themselves as they travel the road of life themselves.

Later, when they reach another stepping stone (or level, one might say) of life and have learned new techniques or methods necessary for human growth, they will again share these teachings/beliefs with others. Other beings then start to merge their way of thinking to include these new nuances, each spreading their own trails during a lifetime.

You can start to see how one's upleveled belief systems start to influence the lives of others. You can envision how these stepping stones of growth can start to uplevel the lives of humanity, one being at a time, for if all beings adopted these new loving ways of life, the human experience would be quite different than it is at present for most of Earth's inhabitants.

How is this achieved–the upleveling of humanity?

The answer is within each of you beings currently on Earth. If each person does their own small (even a very small) part, the combined result can be momentous and virtually life-changing for everyone.

Think of it like this: if all humanity is walking in the same direction of a change, it seems seamless, completely natural. If, however, there are those resistant to this change or set on another (opposing) view, the path can be much more tumultuous and fraught with arguments and unhealthy actions.

This method would not serve humanity. Instead, we all need to seek peace, compassion, kindness and think of all beings as One. Leaving our individual trails that benefit others is one way to achieve this.

Each doing our own part, no matter how small or seemingly insignificant, can yield a tremendous result. A result so bold and substantial as to change the face of humanity and the human experience.

This is the goal of the Pleiadians and why we work so feverously to attain it using beings such as yourself.

Do not worry that your efforts will be in vain, even if they may seem so at times. Remember all great changes come at a price. The price is often ridicule and ostracization. Pay no matter. Pay no attention, for those who are ready to learn the message will hear it.

They will be tuned into the trails others are leaving behind in their life's journey. They will be ready to pick up the slack as it were and help with this great mission. They will be ready to leave their own trails, thus helping to make this great change in humanity.

Remember: we are all but cogs in the great machinery of life, each intertwined with another, each doing our part, and any missing part will cause the system to fail.

Think of yourselves as essential parts, all you Pleiadian descendants!

Spiritual Well-being

Once again we will address human behavior and what steps or teachings are necessary to adopt in order to advance.

This time we will discuss spirit, your spiritual well-being to be exact. This is especially important because humans tend to neglect their spiritual well-being or health. It can be just as important as one's physical health. Humanity needs to learn how to properly tend to one's spiritual health in order to fully develop as a being.

Neglecting one's spiritual health can lead to dire consequences. We have all seen it happen, even if we didn't realize it at the time. Think of the lonely person who just drudges along doing their occupation and has few if any interactions with other people. They don't tend to a garden, have a pet, or enjoy any hobbies. They spend their non-working hours indoors, never interacting with people, animals, or the outside environment. They may be depressed. At a minimum they seem to be sad. There is no spark of joy or happiness in them.

This is because their spiritual health has declined so much that the spark has gone out! At a minimum, their spark has dimmed significantly. They have no reason to arise in the morning except for the obligation of their boring and unfulfilling job. Their body may become riddled with infection. Their mood is sullen and sour. Basically, they have lost most of their will to live!

Let this be a lesson to every being – Do not let your spirit wither away! Life is meant to be a joyous and triumphant journey even if it has both ups and downs. With the proper understanding we can learn to appreciate even the low parts.

But this sad soul has yet to experience the fullness of human existence because their spiritual light has been extinguished.

Instead, we must learn to identify and acknowledge those beings lacking in spiritual fulfillment and help them along their way. There is no need to shove your beliefs down their throats. Merely serving as an example will make a difference. By observing your cheerful attitude and perhaps sharing a few of the practices you employ to assist in your spiritual upliftment, they will be spurred to do the same.

Sometimes beings wither away just because it never occurred to them that life could be any different. Perhaps they didn't learn this growing up in their family. Perhaps they weren't exposed to others who had learned to grasp life by the horns and fully enjoy the ride.

Your small contribution along with those of others can make all the difference. At any rate, trying to set a good example of how to tend to one's spiritual health is a worthy venture.

Next, we will talk about ways to improve one's spiritual health. Lastly, we'll discuss why this is so very important.

Improving One's Spiritual Health

Improving one's spiritual health is an individual responsibility, but others may assist. Learning to acknowledge the need to nurture one's spiritual health is the first step. Beings must become aware that something is amiss in their life (actually their body.) Either they must be learned in these practices through familial activities or become aware of such feelings on their own.

Those beings raised in a consciously aware family may seem to have a step up in the process, but that is not always the case. Their journey may be compounded with additional or more complex growth challenges to overcome. It's important to keep in mind that each being's journey is slightly different from another's; it's all part of the plan.

Those who must find and learn to correct their spiritual health issues are in the mainstream at the present time on Earth. Typically, they experience some sort of crisis or breakdown necessitating changes in their daily human life. This is how change often takes place in the human life experience. We reach a crisis of sorts, making us aware that change must occur so as to relieve the pain or burden being experienced.

Such activity may take the course of a mental breakdown. Other times it might just be a realization that something is "off," or one feels out of sorts. Most feel something in between these two scenarios. It's a matter of becoming aware of your body and sensing there's a problem that needs to be changed. Knowing or learning that something is amiss in your body is the key factor that spurs individuals to work on their spiritual health.

They may wonder to themselves, "What is wrong? Why don't I feel good? What is missing?" These questions of self-reflection are an excellent method of getting in touch with your body and figuring out what is needed.

Often souls will realize they haven't been exercising daily or as is their usual practice. This lack of movement or being outside is causing lethargy or mild unhappiness in their daily life.

Perhaps they've been too busy to meditate or haven't yet set a practice of daily or regular quiet time for their thoughts or communication with their God. Whatever the crux of the issue may be, self-reflection is key to finding a suitable cure or antidote.

At times, beings must seek outside counsel to solve such issues. It could be a trusted friend or a professional. A good friend who knows you well can often pinpoint the change and suggest an antidote quicker than a medical professional simply because they know you better or for a longer period of time. In times of quandary or more difficult times of declining spiritual health, the help of a professional can be key. Although they may not be schooled in spiritual practices, they can help identify problem areas and possible solutions.

In the end, though, it is each person's sole responsibility to take action to correct or compensate for one's ailing spiritual health. It is a mandate from God for each of us to learn how to tune in to our body and discover what it needs so we can live a happy and prosperous life. It's a tenant of one of our responsibilities for being granted an opportunity to reincarnate here on Earth.

Why It's Important to Improve One's Spiritual Health

Our day begins with sunshine or rain, maybe a cold breeze or even snow. How or why does 'how the day begins' affect our mood, our spiritual health?

The answer: Because we let it. This observance should be a lesson to us all. A lesson we can use to our own benefit.

If sunshine and warm weather make you start the day in a happier mood – then create your own sunshine. Create the type of morning routine or ritual that feels right for you.

If it's a quiet time you crave, give it to yourself first. Create the scenario that feeds your soul (your spiritual health) first. Do not delay by telling yourself, you will do it later. The day may get away from you, filled with mundane tasks and other obligations, leaving you drained spiritually.

But if you serve your needs first, you will have the stamina to pursue other activities seamlessly.

How does this work? You see; by meeting the needs of our own spiritual health, we feel more fulfilled, ready for the day and happier in our soul. This newfound energy can help propel us through our day to accomplish those tasks we want or need to do.

How one starts off the day has a tremendous effect on the remainder of the day.

Think of this scenario: You wake up to a blaring alarm. Still dizzy and not yet awake, you push the snooze button. After another minute of slumber, followed by the repeated buzz, you reluctantly arise. As you stumble out of bed, you stub your toe on the bed frame, eliciting a cry from your mouth. You splash cold water on your face and try to wake up.

Now think about this scenario: You wake up without an alarm. You hear birds chirping outside your open window as a gentle breeze makes the curtains waiver. You sit up in bed, collecting your thoughts and feel an immense amount of gratitude for having another day to experience life here on Earth.

Which scenario seems preferable to you? It's not that one scenario is better than another. Another completely different scenario might be your preference. The key is finding a routine that feels right within your soul, one that makes you happy to be alive, a method that fuels your spiritual health.

This is a necessity for all beings to acquire or learn how to acquire if they are to have a joyous life. It is theirs for the taking. It is their birthright. However, one must learn how to create such a life.

Learning to develop one's own morning ritual can be the first step to creating one's perfect life. Once you decide what feels good to you and what you want to experience, then go about creating such a system or routine.

After you've nailed down the morning routine that feeds your soul and thus impacts your spiritual well-being, then expand your vision to your daytime actions. Later focus on the type of life you'd imagine if only you could. Once you take steps to creating the type of human experience you desire, your spiritual health will flourish. You will have done it – mastered this thing called Life.

You will feel happy, joyous, and grateful inside. It matters not if you meet resistance or trials, your soul will hold and keep you safe. Your temperament will not change. You'll be grateful for a wonderful life, even with the challenges. You'll realize the tough times are there to teach you lessons and help you grow. Your spiritual well-being will grow as you learn to embrace these less-memorable components of life.

This my dear, is how it's done. This is how humans learn to live an extraordinary life–a life filled with their own happiness and a life lived by example to others. The type of life that helps lift the consciousness or awareness of all beings. This helps all of humankind evolve. Such a life is key if we the Pleiadians wish to impact humanity. We all must do our part.

So each day, as you begin, think about nurturing your own spiritual well-being first. Design a life that feeds your soul and fuels your purpose. It's your birthright and your responsibility to humankind!

Be well. Be safe. Be content with your journey.

What are Miracles?

Continuing on this path of growth and enlightenment, next we will tackle the occurrence of miracles.

Let's be plain here; miracles are not miracles at all. They just seem like miracles because they are such a happy occurrence.

Actually, a miracle is the manifestation of whatever a being creates. It's created first in the mind-the conscious mind. This triggers behaviors within the sub-conscious mind to make your wish or wishes occur. Often, the help of others is elicited at a subconscious level.

Because we are all eternally connected, our thoughts intertwine at times. This is how another being seemingly assists in the manifestation of your dreams, thus creating what humans deem a miracle.

At times, these instances are quite complicated and interwoven. Things occur that appear to be totally unrelated, but in the end, the wish of one party (and possibly more) culminates in a situation deemed a miracle.

We often go through life wanting and desiring a certain outcome but suddenly have an unrelated outcome. It might be shocking, debilitating, or even painful, but in the end if it's your true desire, circumstances will again change and eventually lead to the originally-desired result.

We've all seen, heard, or experienced events that would be deemed unfortunate, only to have these great low points in life turn into life's high points, greatest achievement, or our heart's desire.

This is how it works. Our conscious mind develops the idea, plants the seed as it were. Our subconscious mind is then the taskmaster, taking minute steps to achieve this goal, often eliciting the unconscious help of others.

Later, we will examine and explain more about the unseen connectedness we all have with each other. For now, it's just important to understand that one's goals or desires are not achieved in a vacuum consisting of only your conscious and subconscious mind. Often such miracles can only come to fruition with the unseen assistance of others.

Such is the way of life, for we are all involved in the daily creation of miracles in the lives of others.

Greatness

All beings can achieve greatness. In fact, we are all made or created for greatness. Whether we achieve it in our lifetimes is up to us, but we are all quite capable.

This is because we were all designed by God, the great one, source, or our own individual higher power, to innately have greatness. It's much like the bud of a bloom of a flower. All the potential is there. Provided the conditions are right – sunshine, water, and proper cultivation – the bloom will expand or grow into a beautiful flower.

Without sunshine, the plant will wither. Without water, the plant will shrivel up and dry, causing it to die. Without cultivation, it will also die.

But what exactly is the cultivation part? This, my friends, is exactly what is missing in the lives of so many humans. It's the spark. It's the tending to by others that nurtures one to grow.

Much as plants and trees often need others (be it other botanical plants, humans, or forest animals) to grow to greatness, humans need this also.

You see, we do not exist in a vacuum. We need other life forms to reach our potential.

A tree needs sunshine and water that is for certain. But it also needs animals to interact with it, like squirrels climbing or birds singing in it. Trees often communicate with other trees, forming large, intertwined root systems. Just like those root systems are hidden beneath the soil, our attachments to other people, animals and nature might not be visible; however these often-unseen ties are what sustain us or nurture us during the storms of life.

For example, remember when you felt a bit blue and decided to take a walk outside? Being among Mother Nature enjoying the trees, flowers, and chirping birds, you are often able to clear your mind and reach a higher level of happiness, pushing your sad feelings aside.

This is because we (meaning all living entities) are all connected, and we rely on each other for nurturing and growth.

This nurturing is essential for our growth to expand to the levels of greatness!

Humans should regularly walk in nature and fuel the bellies of the environment, much like the outside elements will fuel you, so all can achieve greatness.

You see, man cannot achieve greatness (his or her great milestones) in life without the help of other beings.

We must learn to rely more on others, not as a crutch, but as an added, although vitally important element to our well-being.

Let's take a person with moderate capabilities in a specific field. They have a good ability to perform a certain task, whether it be in engineering, writing, or designing new equipment. However, if no one comes along to offer support and encouragement, it is doubtful they can master their trade adequately to reach mastery or as one would say, "greatness." We all need the assistance from others, even if it is a little push to spur us to greatness.

This greatness is but another example of how all creatures on Earth are interconnected, the crux of the human experience. Pleiadians are here to assist in this important endeavor.

The Next Phase

The next phase of human development is upon us now. It may take years for it to be completed, but it will be completed nonetheless with the help of many Pleiadian and non-Pleiadian beings. Since we are all One unity, it will be a joint achievement. More about that later.

How will this upleveling occur and what does it mean?

Each being, a human being in this case, has their own individual job or task to do. It's much like an army of ants, each doing their part to literally move a mountain.

We each have a task to accomplish in our lifetimes. When we do our part, humanity reaches closer to our goal. It takes multiple lifetimes of struggle and strife to accomplish this goal. Beings must learn to be patient to achieve this end. This is but one of many lessons we each are to learn.

The timeline or when this great task is accomplished is not the end goal, rather it is the joint accomplishment of reaching new heights as a collective of human beings that is the crux of the matter. Joint action toward a common goal is another important lesson our lives take with us.

It is this joint cumulative ongoing persistent activity that exhibits that we are indeed One being that will help set us free on our human journey.

This journey takes lifetimes as we mentioned before. It will be uncomfortable at times. Many will give up and need to try again. Others will persist and make progress sooner or seemingly faster. Since time is virtually standing still in a vacuum while this process is unfolding, the length of time required to achieve or learn the necessary lessons is not really that crucial.

What is important though is that each being try his or her best each lifetime and make some small inkling of progress so that eventually all humanity rises together to new heights of understanding, exhibited by their acts of love toward each other and the rest of the Earthly environment. This includes animals, insects, and the physical environment such as the sky and bodies of water. It is this co-existing of all nature and all living beings that represents our journey to understanding that:

We Are All One, one collective being, dependent on and affected by other segments. That will be our greatest accomplishment.

Knowing in our soul that we exist as One being is the greatest lesson we are here to learn.

Do not give up my dears. Instead give it your best attempt. Work towards this goal each lifetime.

The Pleiadian masters urge all being to focus their daily efforts towards achieving this lofty goal, for it will culminate in a happier and more joyous experience for all as a result of this new found understanding and cultivation of love.

We are all love, created from the divine. In time, when all are ready to ascend, we will return to love and unite with our deity. Such a glorious sight and one truly worth all your effort.

The Pleiadian Masters Ozo, O'Berta, Marietta, and Dono plus the others before them who paved the way.

The end.

Bonus Book Excerpt:

The New I AM Document, Volume I

Lesson 1 - The Purpose

In the beginning, God created the Heavens and Earth as a playground, much like a classroom for mankind to learn and grow, to become all he or she can be.

The purpose was the growth or Ascension of mankind to the next level of evolution where all men/women are all-knowing and very much aware of their connectedness to each other and to the Earth—to include the environment and all the animals. We are all truly connected as One; much, much more than the individual man knows at this moment.

All mankind has different issues to handle or solve. We must work together to solve these issues consciously and subconsciously. Part of what we need to learn is how to work together—as One. The point of this entire exercise is that we are One, so we must learn to act as one cohesive unit. All humankind is One. One with each other and One with God.

It's the most important lesson we are here to learn.

With the help of you and many others, God's plan is to raise the planet's shared consciousness to a new level, never known before. A level where we all can experience the glory of an earthly life well before we ascend to the Heavens and someday to another solar system (an alternate Universe).

It will be much the same as this one—a playground for our enjoyment and eventual growth. This is a repeating pattern in the Universe. One that humankind is not yet aware of. With your help and that of our other Ascended Masters though, someday all humankind will know and experience this type of monumental growth.

Lesson 2 - God's Plan

God has a plan unknown to most of humanity.

A plan in which He will lift up man to new heights, new teachings, but only if they search and try for answers. They must ASK. It's the key. No asking, then no spurring onto new thoughts, new resources and new advancements.

We all have this capability, but few try and therefore few take advantage of it.

We all should know by now that there is more to life than what we experience. We can feel or sense things, but we must pay attention. We must be aware. Aware of ourselves, our own body, and the world around us.

One who pays attention can see his relationship with the outside world. One who does not open his eyes to this reality will remain stuck to the old ways.

Those who can see need to be the light (or the eyes, as it were) for the others—those individuals blind to outside occurrences.

It's imperative we help each other, lift each other up, light the way to move humanity forward.

This is done in a myriad of ways.

Encouragement is one such way. It is a kind and loving method used to spur another to greatness, to help them be all they can be. It's the mastery of the method of encouragement that will lead this planet forward. We must all do our part.

Another method is Empathy. We must learn to exhibit empathy for the plight or path of others. Not judging, for that alone is reserved for God and even then, God's only requirement is that we each try our very best, without limits, for then we know (and God knows), our best effort was given.

There are no regrets if you give your best and come in second or last. You know in your heart you have done the best to your ability.

Back to empathy. The kindness and sorrow we feel for others are important to comprehend. No man will advance if he cannot feel the pain and sorrow of another.

Remember we are all connected.

To deny the feelings of another is to deny yourself a part of yourself. This is not what God wants; He prefers we treat each other with kindness as if it was us with whatever affliction, because it truly is us, a part of us, with the here-to-for-mentioned affliction. Important lesson to learn.

Another trait we are to accept is the practice of Gratitude-giving. This doesn't have to be a daily list, although that practice will help many develop the habit, but rather a continuous appreciation for all one has, is capable of, and for the wonderment of our beautiful planet.

Living a grateful life is an on-going state of being, not something to complete merely because it's on your "to do" list.

It's truly a way of being, something that permeates through one's body, which comes straight from the core of a grateful heart.

Lesson 3 - Playgrounds

The world continues to be a playground for humanity to learn lessons and it will be so until such time as all become kindred spirits leading towards Ascension.

When all have learned all the lessons, they will ascend to the heavens above and God will destroy their beautiful home, the Earth. It will exist no more.

All beings, upon their turns, will eventually ascend to another, even better/brighter solar system, where God will again prepare a wonderful playground for their enjoyment and growth. This new place will be called Taos—New World.

Beings will appear (look) differently, without so many features, for they will have learned that physical features do not define a person. It will be the heart or the spirit or even the soul inside each of us that will be the pertinent part of our beings. This is easier to accept once we have learned these lessons: outside or exterior ≠ the interior. We all should delve to understand the "interior" of others if we are to learn these lessons and advance or ascend to the next level.

Future lessons to learn involve acting with even more compassion and love for others at a level that even the most advanced beings at present cannot comprehend. Yes, there is much more to learn and there will continue to be so.

Humankind or extraterrestrial beings as we really are, actually have many more lessons to learn. This will take eons to discover, learn and absorb into our souls.

Lesson 4 - The Process

We all have things we don't know or understand.

Being on this journey helps us to find these answers. All we have to do is ASK. The answers will come in the most serendipitous ways. It will seem like happenstance, but it is all carefully planned out by God Himself, as much as our life is planned.

We all have free will to do as we choose, but the lessons we are here to learn are all set in stone before we are to arrive.

You see, how are we to learn and grow if we do not learn the lessons? Yes, lessons can be painful, yet we should be thankful, for without the lesson there is no growth, there is no change. Learn to accept these challenges or experiences as a part of life. It's part of living your best life.

Failure to learn means one merely has to repeat the lesson. Reason it is so important to pay attention.

We all can do better in life, but in order to do better, we must know better. To know better, we must learn and grow. Experiences or challenges on Earth is the way God teaches us these lessons. This is the reason being allowed to experience life on Earth is such a cherished experience.

Hard learners, or students most interested in advancing, are chosen first and then allowed repeat experiences. It's quite a competitive or shall we say a "sought-after" experience.

When one's time comes to advance to an earthly visit, proper preparations are made. The memory is wiped clean, except for some of the lessons. The reason is to avoid confusion of past lives. You wouldn't want to be living an experience or life in one time and culture and become confused with an earlier situation.

However, strong connections with some people, your clan, or connectors, is still maintained through feelings. You are able to feel some sort of connection. This feeling transcends lifetime experiences. The reason is that as we advance and become more in touch with our feelings, we will experience a "knowing" or sense of ease when we first encounter one of these beings.

Such is the case for Soulmates and even more so for Twin Flames. The connection is so strong, one cannot ignore it. One might not understand it at the time, but nevertheless, it cannot be ignored.

Lesson 5 - The Cornerstones

Kindness, Faith, Love and Trust are the cornerstones for building a good life.

Man needs all four to lead a productive and essential life.

Kindness teaches us about others. When we are self-centered and self-serving, and not showing or feeling kindness to others we will not thrive or grow.

Kindness teaches us many lessons, one of which is to put the well-being of another or others before ourselves. The miracle is that by doing so, we actually help ourselves. We grow. We learn to be compassionate towards others. We treat others as God would want. And it fills a need deep inside our souls. A need we cannot understand or fathom until we experience it.

Faith is another necessary component, one that will carry us a long way. A long way as in over a long period of time, but also a long way in the sense of soothing or comforting our souls. In times of need and despair, it will carry us through. The lesson learned, among others, is that by having Faith in God and His ways, we will be carried through our trauma (or opportunity for growth, as we often refer to such challenges).

Do not be afraid to ask God for help and accept His help. This is part of the process of Faith. It also feeds and soothes our inner soul.

Love is by far the most important, the deepest and longest-lasting feeling we are here on Earth to experience. To experience a life of deep and abiding love is to experience one, if not the most, deeply divine experiences on Earth. The love stays in your heart through multiple lifetimes. You may not always understand it, but you will always feel it.

The most advanced masters are so filled with love that it encompasses their whole or entire being.

Love is where the "light" comes from. Those advanced in feeling love as a continuous way of life put off a glow—a light—that others can almost see. In fact some can see it. Nonetheless, others can certainly feel it.

This light inspires others even if they aren't aware of what is happening. They will tend to follow or be drawn to others "with the light." This is how they will learn and grow. This is the reason it's important to always set a good example...others tend to always observe. If they "catch you" doing good, it spurs something inside of them to do the same. They aren't even always aware of this as it can be their subconscious leading them.

Trust is also a necessary component for one's development and growth. We all must learn to trust God for He and only He has complete control over our Universe. We control our growth (or lack thereof), but He controls the classroom, as it were.

The way to build Trust is to have it. The way to have it, is to Trust. It's a never-ending circle or cycle. You just have to jump in like a merry-go-round. Start wherever you are and take that leap of faith.

Trust has a way to calm our inner souls. If we Trust in God, we feel a calmness not experienced in other ways. It's necessary to get to this place of calmness in order for us to grow.

When we are calm, we are kinder to others. It also helps to maintain or attain the proper vibrations for our lives to work perfectly together. Trust helps get you there.

Once you learn to Trust in God, your life begins to "flow." This is the Heavenly Flow of Life that God wishes for all humanity. One of the important lessons we are all here to learn.

Peace and Joy follow when one learns to Trust. Trust is also linked to Love. Greater Love can be received if we Trust it will come in God's timing.

About the Author

Photo Credit: Blanca Madrigal

Hi! I am Janie Jurkovich, known as Janie J, to avoid misspellings and mispronunciations. I started writing later in life when my life took a sudden turn.

In my early sixties, my spouse of 35 years suddenly departed in what might be called a late mid-life crisis. I was not equipped to 'start over' but I had no choice. There was one thing I knew for certain—I was not going to live the rest of my life in the same manner! I was tired of working multiple jobs and doing all the house and yard work. In addition, there was no time for travel, entertainment, hanging out with friends or doing anything remotely resembling 'fun.'

I knew something had to change and it was me! I set about trying to figure out what my new life would look like and how I would get there. The result was my first book, "Live the Life You Have Imagined." Another book followed that depicted my life as a single woman navigating life on my own, titled, "Single and Sixty."

Then something unimaginable happened to me. I had a spiritual awakening and realized I was a spirit medium able to receive messages from beings in another dimension. I described myself as a "spiritually starved workaholic" who never had time to even ponder the meaning of life. This resulted in yet another book, "The Unimagined Awakening."

Then the big news came in the form of discovering my life's mission—to transcribe three volumes of words of wisdom from Archangels! The first volume, "The New I AM Document, Vol. I" is available now. Volumes II and III are in progress.

I am currently living my dream life in Central California where I reside in the country with a flock of sheep, a brood of hens, two dogs, and my soulmate. It is here in this beautiful and serene setting that I have received the solace to discover and live my best life. I spend my free time traveling the globe and teaching others the lessons

I have learned. I continue to write and have multiple books in the works.

My mission is to help others discover what it takes to reach this level of consciousness and understanding so they too can live the life they have always imagined.

For more information visit:

www.TheNewIAmMovement.com

www.JanieJ.net

YouTube: Janie J Soul Connector

www.youtube.com/@janiej-soulconnector

Other Books by Janie Jurkovich

Live the Life You Have Imagined! Simple Ways to Begin Living Your Best Life

Live the Life You Have Imagined: Companion Journal

Single and Sixty: A Reflective and Sometimes Humorous Journey of One Woman's Quest to Deal with Divorce Later in Life

Ignite Your Adventurous Spirit (Contributor)

The Unimagined Awakening: My Journey from Spiritually Starved Workaholic to Connecting to "All That Is"

The New I AM Document, Volume I

Buy Janie's books here: www.janiej.net/store/

Learn Janie J's Connect to Transform™ process to uplevel your spiritual evolution.

When Janie J first started her spiritual journey, she didn't know what meditation was or how to do it. A friend encouraged her to 'just breathe.' From there, she developed her own way to quickly achieve inner peace and calmness in her stressful life. Eventually it took less than a minute. Whenever she wishes to return to this peaceful state, she'll just follow this simple process.

This process will work for you too!

Janie J has prepared an easy to follow, step by step guide to help you "connect" so you too can transform your life, from the inside out!

Use this process to:

- learn and teach your body and mind how to calm down.
- connect to your higher self
- connect to your spirit guides
- connect to God or your higher power
- connect to whomever, whenever you want
- connect to animals and nature

It just takes practice, as does anything in life.

Using the Connect to Transform™ Process will help you on your spiritual journey.

Get the Connect to Transform ™ Process

Free Instant Download Online at:

www.TheNewIamMovement.com/transform

Learn how to connect to your higher self, others, your Spirit Guides, Ascended Masters, the Pleiadians, God, your higher power, Universal Intelligence, and All That Is.

Join the New I AM Movement

Are you ready to change your life for the better?

Are you ready to upgrade your existence?

Are you ready to help move humanity to a higher level of consciousness,

more in line with what our Creator envisioned?

It all starts with YOU!

If you are ready to take the next step on your spiritual journey, find out

where to learn these lessons so you can grow into a better person

and become a more evolved spiritual being.

Go to: www.thenewiammovement.com/join

Just imagine being a part of this extraordinary movement—a movement to move humanity ahead to new, higher levels of understanding and consciousness. Such a noble feat is quite possible if we work together to make this happen!

Join The New I AM Movement as you work on your own spiritual development. Connect to me, to others, your higher self and with your Spirit Guides, Ascended Masters, and Universal Intelligence. We are all available to help you on your spiritual journey.

Made in the USA
Monee, IL
04 November 2026

67386588R00049